THATCHING

THATCHING
A handbook

Nicolas Hall

Practical
ACTION
PUBLISHING

Practical Action Publishing Ltd
25 Albert Street, Rugby, CV21 2SD, Warwickshire, UK
www.practicalactionpublishing.com

© Intermediate Technology Publications 1988

Reprinted 1991, 1992, 2002

ISBN 10: 1853390607
ISBN 13 Paperback: 9781853390609
ISBN Library Ebook: 9781780444352
Book DOI: http://dx.doi.org/10.3362/9781780444352

A catalogue record for this book is available from the British Library.

The authors, contributors and/or editors have asserted their rights under
the Copyright Designs and Patents Act 1988 to be identified as authors of
their respective contributions.

Since 1974, Practical Action Publishing has published and disseminated
books and information in support of international development work
throughout the world. Practical Action Publishing is a trading name
of Practical Action Publishing Ltd (Company Reg. No. 1159018), the
wholly owned publishing company of Practical Action. Practical Action
Publishing trades only in support of its parent charity objectives and any
profits are covenanted back to Practical Action (Charity Reg. No.
247257, Group VAT Registration No. 880 9924 76).

Cover illustration by Satoshi Kitamura
Text illustrations by Ethan Danielson

CONTENTS

ILLUSTRATIONS

PREFACE

This handbook on thatching has two primary objectives. Firstly it is a guide to good quality thatching, describing in words and pictures how to achieve the maximum possible roof life using either cultivated or naturally occurring materials. Secondly it details the limitations of thatch — fire in particular — and what can be done to minimize these shortcomings.

Rural builders, architects, planners and surveyors should find it a useful source of information on the advantages and limitations of thatch. Whilst acknowledging that it should not be used in densely built-up urban areas, it is equally clear that the material has considerable and often unrecognized potential for modern, permanent-type housing in rural and semi-urban locations. Apart from serving as a weather-tight and durable roof covering, thatch also has many secondary advantages.

It uses renewable natural resources whose production and processing can be a valuable rural income source. Thatching is labour- rather than capital-intensive work. It is also skilled work — a professional thatcher can derive great satisfaction from his work and expect respect from the community he serves. And thatch, unlike tiles and metal sheet coverings, can be laid on relatively inexpensive pole timber roof structures.

In using this handbook the reader must appreciate that good quality thatch is more expensive than most traditional types of thatch. To do it well means using more material, more time, and careful craftsmanship. And of course it is pointless to put a high quality roof on a temporary building.

Michael Parkes
Manager, Building Materials Programme
ITDG

1. THATCH — A MODERN ROOFING MATERIAL

Definition

Thatch is a roof covering made of dead plant material — other than wood. Grasses and palm leaves are the most widely used materials; many others, such as seaweed, herbaceous fibres and large leaves provide roofing material for traditional building in some countries.

Traditional Thatch

Historically speaking, thatch, in conjunction with rough pole timber, is the original building material. In the rural areas of Africa and Asia it is often the only available economic roof covering. Elsewhere, it has largely been replaced by rigid inorganic materials like stone tiles, and manufactured products such as metal sheeting and clay and cement-based tiles. However, thatch is still used in the industrially sophisticated and wealthy countries of N.W. Europe because it is pleasing to look at, and is a weather-tight and durable covering.

In many parts of the world traditional types of thatch are not favoured. This is partly because thatch is associated with poverty and not thought of as durable and suitable for modern buildings; but it is also because traditional types of thatch often do have very real shortcomings — they leak, harbour insects, tend to catch fire, and do not last long.

These limitations do not of course apply only to thatch; the vernacular rural buildings of most societies were frequently not expected or intended to last more than a few years. Durability was a lower priority than convenience, low cost and simplicity. Today, however, more and more people require durable shelter and are thus naturally inclined to abandon traditional solutions in favour of new materials and building techniques. The first thing to be replaced is often the thatch.

It is clear that some thatching materials and the techniques associated with them cannot be upgraded to meet today's expectations and improved building standards. Palm leaf thatch, for example, can be improved, but nothing can be done to make it last more than 15 years at the very most.

In contrast, thatching with certain types of grass offers much more potential. Species of grass with tall, rigid stems can last 50 years or more if they are carefully and skilfully laid. So thatch can be as permanent a covering as corrugated iron, or tiles. This handbook describes each aspect of modern durable thatching.

The Advantages of Thatch

Thatching is a modern craft, although it has very ancient origins. Like any living technology it has been constantly developed and improved. Today, thatching can have most of the advantages of any modern roofing material, as well as the original good points which have made it so useful throughout the centuries.

First, it uses locally produced material. Supplies can be increased at very little cost to the community and without risk of upsetting the local economy. Furthermore, it does not need sophisticated machinery or tools; therefore it will

1

not eat into scarce foreign exchange, or use up valuable local capital resources in the same way as, say, corrugated iron or cement tiles. Grass or palm leaves grow again every year and with the simplest care will provide abundant thatching material.

Thatching can provide the basis for small business where people have few or no resources, and can thus increase prosperity and independence. Farmers and thatching craftsmen would both benefit directly and in turn help the local economy.

In most societies thatching is not an alien technology. It is much simpler and less damaging to a community to suggest ways of improving an old craft than to introduce a completely new technology. It is true that a new material may be welcomed at first — but the side effects of sudden change can be extremely damaging. All too often, the new — and expensive — imports do not live up to expectations when they are employed in unfamiliar situations and by craftsmen who are not used to them.

Thatching is creative and satisfying work. It is a craft, and the craftsmen can get pleasure and status from turning a pile of grass into something both useful and good to look at. Additionally, of course, as the thatcher improves, so does his income.

Thatch is a very versatile roofing material and it will keep a house cool in summer and warm in winter. It provides excellent insulation against extremes of temperature. You would need to fix roughly 200mm of fibre-glass matting under corrugated iron to get the same insulation value — 'U' value — as 300mm-thick new grass thatch. Thatch also provides very good sound insulation. Finally, thatch only requires simple and inexpensive maintenance.

Even the best quality thatch does however have some drawbacks.
—It is not suitable for high density urban housing because it is combustible (although much can be done, with minimal expense, to reduce the fire risk).
—Thatching materials are bulky, so transportation is likely to be expensive.
—It involves labour-intensive work, so it will be expensive where labour rates are high.
—Thatched roofs cannot easily be used to collect rainwater.

Conclusion

Although thatch has a lot of advantages the most difficult hurdle is often the reluctance of many people to use what they see as an old-fashioned, primitive material. Perhaps the best way to overcome such prejudice is to show how well thatch performs in practice. This handbook looks at ways to produce good quality, long-lasting thatch that is nonetheless reasonably simple to lay and to maintain.

2. THATCHING MATERIALS

This chapter looks at the different types of thatching material, defines the characteristics that thatchers need and prefer, and describes the sources, the harvesting and processing of these materials. It looks also at the reasons why thatch decays. It concentrates on grasses, as these can be durable and are the most likely to be used in modern building.

Hundreds of different plants are suitable; which one is actually used depends on local availability and the certainty of adequate supply. Fundamentally, they can be divided into three groups. First, there are palm leaves; these have either fan-shaped or feather-shaped leaves. Secondly, there are grasses which have soft stems, and thirdly there are stiff-stemmed, reed-like grasses. Each plant species needs a different type of processing, and a particular way of attaching it to a roof to produce a weather-tight covering.

Sources and Types of Materials

Thatch may come from three different sources. First, from naturally occurring indigenous vegetation, secondly as a by-product of food or cash-crop agriculture, and thirdly through the cultivation of a plant grown specifically for thatching. The first two sources traditionally account for the bulk of the materials used, but carefully planned and strictly managed cultivation is likely to produce the most durable and economical material.

Palm leaves usually come from wild trees or as a plantation by-product, although the sago palm is grown primarily for its leaves by some Pacific island communities. Indigenous palms are often of such vital economic importance that their active protection puts them into the same category as cultivated species. However, palm leaf thatch technology has not been developed in any society to produce roofs with a durability exceeding 10 years, and most types last less than three years.

Grasses, of which several types supply material for roofs which can last more than 50 years, are found all over the world and form one of the largest flowering plant families, with some 9,000 species. Most are perennials, though many propagate annually through the means of self-fertilized seeds. They are economically important as pasture, cereal and food crops, and as the basis of a diversity of industrial and construction materials besides their use as thatch. Cereal straw, as a crop by-product, finds wide useage — wheat and rye in Europe, rice in Asia and millet and sorghum in Africa. On the other hand there are some uncultivated wild grasses and a few specially managed types which have primary economic value as thatch. The most durable is *Phragmites australis* (water reed), which can last up to a hundred years when correctly used. It is mostly grown under controlled conditions in wetlands in Europe but it can also be harvested from wild stands in Africa and Asia.

There is little value in listing all the grasses that are used for thatching. There are too many of them, and the same species may or may not be good for thatching depending upon whether the growing conditions are favourable. Of more value is a list of the stem characteristics that are necessary for durable thatching.

Grass Thatch: Materials Specification

1. Length: Minimum 1 metre, maximum 2.5m
For ease of handling and durability grass with an average length of 1.6m is optimum, but a small percentage of shorter grass is also required for the eaves, gables and top layers of thatch.

2. Stem (or culm) diameter: 5 to 10mm at the cut end
Thatchers prefer the thinner grass as it packs more tightly and can therefore produce a more durable coat of thatch.

3. Straight stems
A bend, usually occurring naturally only at the bottom stem node means that the material will be difficult to 'dress' into place and will also hinder dense, tightly packed thatching work.

4. Strength
There is no scientific measure of strength designed to test thatch quality. Grass is usually tested by crushing the butt between the fingers, experience being the guide to this characteristic.

5. Flexibility
The stem must not be so brittle when dry as to break when being laid and worked into place. Once again no measurement of flexibility is applied other than experience.

6. Leafless stem after combing
Leaves (and weed stems/leaves) left on the culms make grass difficult to handle. They absorb water, look untidy and hamper the thatcher in the task of laying a compact thickness of thatch.

7. Tapering to ear
This characteristic is important for compact thatch laying.

8. Hollowness of stems
Solid pith stems are thought to reduce durability by facilitating water uptake into the body of the thatch by capilliary action. This is neither proven nor disproven, but if choice between materials is otherwise equal, hollow stems should be selected.

Many ·types of grass have these characteristics in nature, but careful management of the growing and harvesting conditions, and correct processing, can enhance the thatching quality of a wild grass.

Material Production

The aim is to produce or acquire, in the least expensive manner, grass which can then be easily harvested in large quantities. It should also have all the characteristics listed above, with minimal processing. The grass should preferably be grown on marginal land so that more productuctive land may be kept for food production.

Water reed (*Phragmites australis*) is the grass most commonly used for thatching material in Europe primarily because it is durable, but also because it grows in dense stands in marshy areas where other crops are not viable. Marsh water levels are regulated by sluices (removable barriers) to produce conditions that favour the dominance of reed rather than competing aquatic plants. Since artificial nutrients are not needed, the only other management is regular harvesting, which is done either annually or biennially during the winter months after the frosts have loosened leaf growth. Cutting is done by machine, and the stems are cleaned with a hand-held side rake before being bundled ready for the thatcher.

Winter wheat is also grown specifically for thatching straw, though the grain is a significant by-product. Cultivation differs from grain production in that tall straw varieties are planted. The straw is less susceptible to fungal attack if less artificial fertilizer is used, and weeding is therefore a major part of cultivation. The straw is cut with a reciprocating blade machine that also binds the sheaves, and it is then cleaned to remove leaves by a revolving drum device that does not damage the straw. The same machine separates the grain from the ears.

In some places, the annual cutting of perennial grasses produces an adequate supply of thatch, but in others it is necessary to manipulate the growing regime by various methods to satisfy local demand. Early growth may have to be protected from the ravages of foraging livestock, or favoured species may be encouraged to dominate and be more productive if the area is burnt annually after harvesting the thatching grass. Local circumstances dictate the appropriate approach.

Where circumstances allow, wild grasses may be cultivated under controlled conditions. A planting programme of seed gathered from the wild should ideally derive from the results of small-scale trial germination and cultivation tests. Most species will benefit from irrigation and certainly from regular weeding.

Harvesting and Processing

The primary objective of a harvesting programme is to gather in mature grass stems as efficiently as is possible, but consideration should also be given to the viability of next year's crop. Perennial grasses will naturally reproduce, but self-seeded annuals should only be cut after the ripe seed has dispersed naturally.

Harvesting can be done by hand, or mechanically. Sickles or scythes may be used

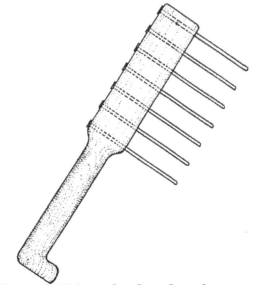

Fig. 2:1 Side-rake for cleaning grass.

5

for manual cutting, or a variety of different machines are available to hasten the task. Fully mechanized harvesting will include tying the grass into bundles; machines that do this are expensive and only justified where the grass grows in large areas in dense stands, and where the market for thatching materials is predictable.

Processing. Under ideal circumstances, harvested grass will be ready for the thatcher straight from the field, but usually the quality of the crop will be greatly improved by processing. Two distinct tasks are involved.

First, the grass must be cleaned to remove leaves from the stems, and to clear superfluous weed growth. This may be done in the field by combing handfuls of grass with a rake at the same time as it is cut. Fig. 2:1 shows a hand-held side-rake. This may be mounted on a waist-height trestle if the work is to be done later, off the field. Alternatively a manually or mechanically rotated combing drum can be used. A standard 44-gallon oil drum mounted in a frame with 75mm-long steel teeth welded to it (Fig. 2:2) serves the purpose well, and can be inexpensively made in a metal workshop.

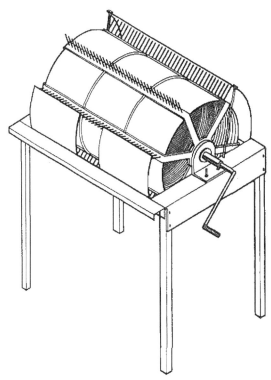

Fig. 2:2 Combing drum.

After cleaning, the grass is tied into bundles. Experience has shown that these should preferably measure 550mm in circumference at the binding, which is tied about 300mm from the cut end. An extra tie may be needed if the grass is particularly long or if it has to be handled many times between harvesting and using on the roof.

Finally it is important that the bundles are kept in dry and well-ventilated conditions until they are used.

Thatch Decay

Thatching grass lasts best if it is kept dry and out of the sun. This is true both for grass in storage and on a roof.

Researchers in Europe and India have investigated the way that thatch decays, and have found that it deteriorates through the combined processes of physical and biological erosion. Thatching technique, which is described in Chapter 5, is designed to minimize the amount of each grass stem that is exposed on the roof surface to the effects of weathering.

Decay starts when the stem surface is physically damaged. This may be caused by wind, rain and swelling and shrinking through temperature change, or a combination of all of these.

Exposure to sunlight accelerates decay. Ultra violet light, reacting with the lignin content of the surface cells (epidermis) causes the grass stems to split, thus exposing the less strongly bonded internal cells (parenchyma). It was found that fungal attack then causes the most significant damage to the stem structure, and that favourable conditions for fungal activity are moisture content exceeding 20 per cent and temperatures between 20 to 30°C. In other words, the grass structure is destroyed in warm, wet weather conditions.

The thatcher's task is therefore to lay the grass stems so that only a minimum area of each stem is exposed to the weather. They should be steeply sloping and tightly packed so that water runs from tip to tip over the surface rather than penetrating into the thatch. If this can be achieved through careful workmanship consistently over the entire roof, then decay will only occur on the outer surface of the roof. After 50 or more years a 300mm-thick coat of tough water reed, skilfully laid, and exposed to the damp temperate climate in northern Europe (for example) will have eroded away to about 180mm thickness, but should still be weather-tight.

3. BUILDING TYPES AND ROOF STRUCTURES

Thatch is a very effective roofing material provided that it is well laid and that the building is designed to suit the characteristics of thatch. In general, thatch works best if it is fixed to simple roof structures with the same roof angle throughout. Changes of angle within the roof are more difficult to thatch and mean that the thatch will decay at different rates. However, thatch is a flexible covering that can satisfactorily be fixed over an irregular or uneven substructure.

All thatched roofs have common features although each material may require a slightly different structural base. This chapter describes these, and shows which features will be difficult to thatch successfully.

Roof Slope and Thatch Load

The two fundamental aspects of a thatching material are the minimum roof slope it needs (measured in degrees/°), and its weight (measured in kilograms per square metre/kg/m^2).

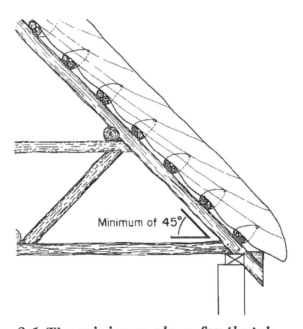

Fig. 3:1 The minimum slope for thatch.

Thatch should be laid on a roof pitch of at least 45°, preferably 50°. This applies to all grass and palm thatch. The steep slope is needed so that water will run off from the roof surface with minimal penetration into the body of the thatch coat. At a pitch lower than 45° the thatch will decay very rapidly (Fig.3:1).

This roof slope requirement means that thatch is most suited to small- and medium-span buildings. Spans wider than 5m require long lengths of rafter timber and carefully designed bracing systems. Suitable roof profiles and bracing are shown in Figure 3:2.

The weight of a new thatch coat depends on the type of material used and the thickness and density of the thatching work. The heaviest material (water reed — *Phragmites australis*) is likely to be about 40kg/m^2 for a 300mm thick coat when dry, but up to 50kg/m^2 when wet. Other grasses suitable for thatching are lighter, usually in the range between 25 to 35 kg/m^2.

Timber dimensions needed to support up to 50kg/m^2 varies with timber variety and the bracing system used to strengthen the structure, but for safety rafters should be 100 x 50mm, and spaced at 750mm centres, with horizontal battens cut from 50 x 25mm timber fixed at approximately 300 mm centres. Bracing timber — collars, ties and diagonal wind bracing — should be 100 x 50mm timber (Fig.3:2).

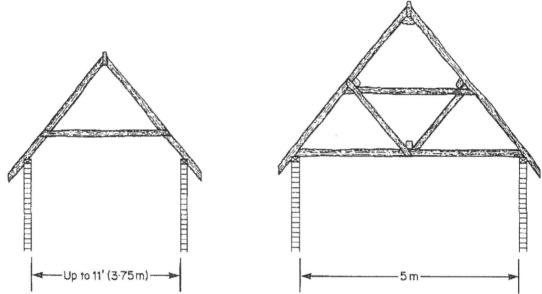

Fig. 3:2 Simple roof trusses for small-span roofs.

Palm thatch is generally lighter (10-20 kg/m^2) because it is laid thinly, so the roof structure need not be so strong.

Thatch does not however need to be laid on sawn timber. It is flexible and will mould itself to the irregularities that occur if pole timber and split wood battens (such as bamboo) are used.

Additionally, unlike roof tiles and rigid corrugated sheets which need straight and precisely spaced roof timbers, thatch is forgiving of distorted or roughly machined timber. In fact, this is one of its advantages — it can be laid on less expensive, unprocessed timber.

Design Detailing

Batten Spacing

The dimensions and spacings of the timber substructure of rafters and bracing members depends on the weight of the thatch material being used. Details are given above, but the spacing of horizontal thatch support (the battens) is directly related to (a) the length of the thatch material, (b) the method of securing the thatch and (c) the thickness of each layer.

(a) Grass thatch is secured in ascending, overlapping courses; the fixing position is approximately half way between the butt ends of the grass (which are exposed on the surface) and the tips of the grass (which are covered by the subsequent

10

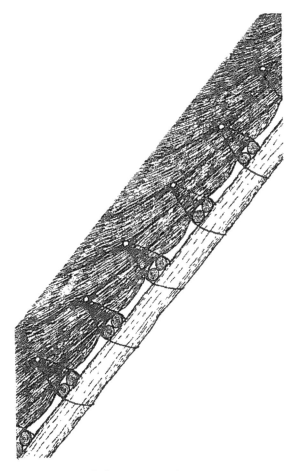

Fig. 3:3 Grass thatch.

course) (Fig.3:3). When using long grass (over 1.5m), which covers more roof area with each layer, the space between the fixings of each course is wider than for short (1 to 1.5m) grass; and so the underlying battens can be further apart.

(b) Thatch is usually stitched to the roof structure (Fig.3:4). Batten spacing will vary with the length of the grass, but there must be sufficient free space beneath the batten to allow a thatching needle, either curved or straight, to pass around the underside of the timber. This is usually not a problem except at the bottom of the roof where ceiling fixing may have to be delayed until the thatch is secured.

When screwed bindings are used the battens must be strong enough to hold the 25mm-long screws. Sawn softwood batten timber is needed. Bamboo is not suitable.

If the thatch is nailed down with crooks, batten timber will generally be too weak to hold the crook. In this case the thatch must be fixed to the rafters, which must therefore be set at the same centres as the horizontal distance between the fixings — usually about 300mm.

(c) On a thinly thatched roof each layer will cover a greater roof area than with a thick coat. Thus the fixings are further apart and consequently the batten spacing can be wider.

Taking these three factors into account, the maximum batten spacing, for grass over 1.5m long would be 350mm. Minimum spacing, suitable for 1m-long grass, laid thickly, would be 250mm.

11

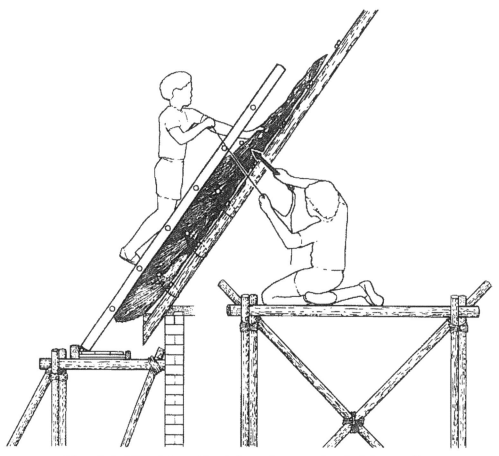

Fig. 3:4 Stitching thatch using a straight needle.

(d) Palm leaf thatch battening is arranged to suit the shape and length of the leaf. Chapter 5 goes into detail about the different palm thatch methods, and hence also the batten requirements.

Eave and Gable Verge Detail
An essential prerequisite of durable grass thatching, and an advantage for palm

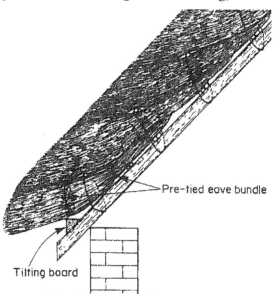

Pre-tied eave bundle

Tilting board

Fig. 3:5 The tilting board.

12

thatching, is correct timbering at the edges of the roof. The eaves layer of thatch should be considered as the foundation for the rest of the roof; it is held securely, by being fixed over a raised tilting board. This can be formed in a number of ways to suit the eaves design of the particular building (Fig. 3:5), but must always provide a bearing surface some 40mm thicker than the subsequent battening. Similarly, at the gable, a fascia board should be fixed so that it stands proud (by 25mm) of the ends of the battens.

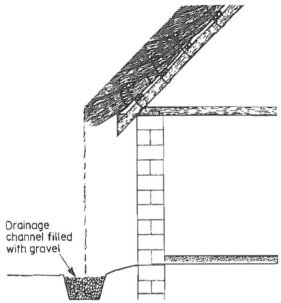

Drainage channel filled with gravel

Fig. 3:6 Ground-level drainage.

Guttering

Guttering is not common on thatched roofs. Rainwater usually drips from the eave and is carried away at ground level in a surface drain or trenched, gravel-filled 'French drain' (Fig. 3:6). Where it is necessary to fit eaves-level guttering this must be made wide enough to catch water not only when the roof is new, but also when the coat has eroded some years later. The gutter should thus be about 225mm wide to allow for this weathering. The gutter needs to be placed with sufficient space beneath the thatch to allow unrestricted access for regular (annual) removal of debris (Fig.3:7).

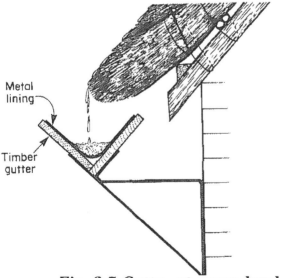

Metal lining

Timber gutter

Fig. 3:7 Gutter at eaves level.

Chimneys, and other projections

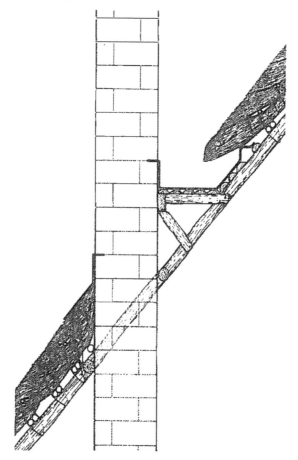

Fig. 3:8 Back-gutter behind chimney.

Projections through the roof such as chimneys and service ventilation pipes should preferably be placed astride or adjacent to the ridge, and if the design allows it, attached to or integrated into the gable wall. Placing them in the roof slope means

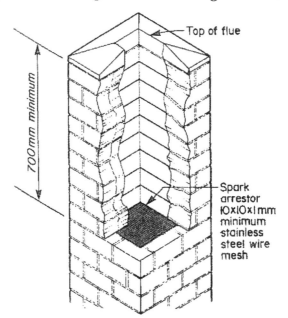

Fig. 3:9 Spark arrestor for chimney flue.

14

that a back gutter must be built and weathered with sheet metal. (See Fig. 3:8.) Design for the flashing at the sides and face of a chimney, preferably with lead, or alternatively with cement mortar, must allow for the thickness of the thatch. As a fire precaution the chimney must be built tall enough to discharge above the highest point of the thatch. As an additional precaution a fine wire mesh spark arrestor should be fitted into the top of the flue (Fig.3:9). All in all, it is safer and easier to thatch a roof without any projections through the roof space.

Ceilings

Ceilings can be fixed in the usual way to horizontal ceiling joists, in which case the joists must be strong enough to support a person who may need access to the inside of the thatch. This could be during re-thatching work, particularly if the thatch is stitched to the rafters.

Fig. 3:10 Fire-check thatch lining/ceiling.

Alternatively the ceiling may be attached directly to the rafters (Fig. 3:10) and thus serve the additional purpose of acting as an internal fire-check lining. A fire resisting sheet lining fixed to the rafters has been found to be the most useful fire precaution with thatch, particularly in temperate regions where the greatest fire risk is from sources, such as kitchen accidents or electrical faults, which are inside the building. However, this sort of lining has to be removed when the thatch is renewed if the fixing method is by stitching to the rafters.

Lightning Conductors

Every well-designed building should have a lightning conductor, and thatched ones are no exception. On thatch however, the conductor must be arranged so that it is separated from the surface of the thatch by a clear gap of some 300mm. The conductor should be carried on wood brackets fixed to the roof structure, as shown in Figure 3:11. Alternatively, protection may be provided by a conductor attached to a pole set next to the building; the conductor must be higher than the top of the building.

15

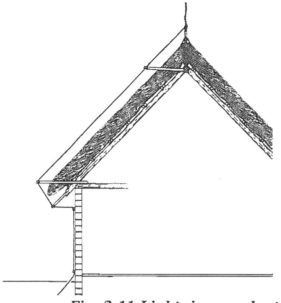

Fig. 3:11 Lightning conductor.

Roof Windows

Dormer windows, set into the roof slope, and 'eyebrow' windows at eave level should, for maximum thatch durability, be avoided (Fig. 3:12). They invariably have a shallower pitch than the rest of the roof, so the thatch over them decays faster. However, since they can be an appealing feature of the building design, they are often included, especially when the otherwise superfluous attic space would be

Fig. 3:12 Dormer and eyebrow windows.

unused. The roof timbering around these windows must be designed to accommodate the thickness of thatch, and should be as steeply sloping as is possible, and never less then 40°. For dormer windows the bottom of the sill must be set at a minimum of 450mm above the structural roof level to accommodate the thatch thickness and suitable flashing.

Valleys

The junction between two adjacent roof slopes is vulnerable to particularly rapid thatch decay as water flow is concentrated in the valley. So although the thatch surface tends to be tighter in a valley (just as it more open at a roof hip), the extra wear invariably leads to the early failure of the valley thatch. One answer to this is to line the valley with tiles or metal sheet and to end each thatch course just short of the valley. This has the dual purpose of simplifying the thatching work and producing a more durable valley covering (Fig. 3:13).

Fig. 3:13 A more durable valley lined with tiles or metal sheet.

4. THATCHING TOOLS

Introduction

Durable thatching depends to some extent on using good quality and carefully prepared raw materials. But even the toughest and straightest water reed can only be weather-tight if it is laid with care and skill. Good workmanship, as in any craft activity, is only possible by using suitable tools. Fortunately, very few specialized thatching tools are needed; none are expensive and those that cannot be bought require only simple blacksmithing and carpentry skills to make.

Tool requirements for palm thatching are fewer than for grass-type thatching. The work involved in maximizing the roof-life of palm leaf thatch is done mostly while harvesting the material and preparing it on the ground. The only essential tools are suitably shaped knives for trimming and splitting the leaves. Work on the roof is largely a matter of laying down and securing — by stitching — panels of prepared thatch. By contrast, the durability of grass thatch depends as much on careful work on the roof as on preliminary preparation of the material. Good grass thatching work is achieved more easily by using the correct types of tool. Thus the tools described here are those needed for grass thatching.

A grass thatcher needs various tools for different aspects of his work. First, to assist with initially placing, and temporarily holding, the grass in place on the roof. Secondly, to help in firmly securing the thatch. Thirdly, to beat it hard into its fixings and simultaneously to produce a smooth and compact exposed surface. Fourthly and finally, to trim the rough edges at the eaves and gables of the roof.

The tools described here are designed for these jobs. However, they may need to be adapted slightly to suit the particular characteristics of the local grass, or perhaps the preference, or the strength and size of the thatcher. Any local modification to the general specifications presented here are a matter of choice rather than of necessity.

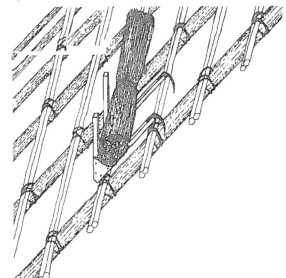

Fig. 4:1 Grass support to hook over roof battens.

19

Laying the Thatch

Bundles of prepared grass are carried up to the roof and placed next to the thatcher to be ready for use. For convenience a small cradle-like support that hooks to the roof structure may be used to prevent these bundles falling back to the ground. This need not be an elaborate device. Figure 4:1 shows one such reed holder.

Temporary Fixing

(a) Two pegs are used to retain the edge of each layer of thatch whilst work proceeds (Fig. 4:2). These are moved along each course as each new bundle of grass is laid. Twenty to thirty pegs will be needed, the number depending on how many courses of thatch are needed to cover the roof.They need be nothing more than simple wooden sticks about 400mm long with a point at one end, or they may be more elaborate blacksmith-made pegs like those illustrated here. One peg should be kept aside for use as a depth gauge to check the thickness of thatch as the work proceeds. It should be marked off in centimetres along its length.

Fig. 4:2 Pegs used to hold the edge of the course whilst working.

Fig. 4:3 Temporary fixing whilst thatching.

(b) Each layer of grass is first laid loosely before being securely fixed, so it is useful (though not essential) to have some means of holding the grass temporarily whilst it is being worked into place. This could comprise a thin, metre-long piece of split bamboo which will be laid horizontally near to the outer edge of the course. It is held in place by several hairpin-like hooks made of sharpened wood driven through to the thatch beneath the course being laid. Alternatively, the bamboo could be held by specially made hooks which, when twisted through 90 degrees, will catch under a roof batten. The illustration (Fig.4:3) explains.

Fixing the Thatch

Many different ways of securing grass thatch have been developed or evolved to suit local requirements and conditions. Here, the most effective ones, and the tools associated with each, are briefly described.

A thatch covering that is laid on a structure of round poles or hardwood rafters is best fixed by stitching, but where sawn softwood timber is used for the structure the thatch can be fixed down with specially made hooked nails, called crooks, or with screwed bindings.

Stitching Where labour is inexpensive and where the interior of the roof space is open for easy access this can be a two-person job, using a straight thatching needle. The thatcher pushes the needle, loaded with string or wire through to the person inside who then loops it around a batten and pushes it back through the thatch to the outside. A 500mm-long wood or metal needle is used, pointed at one end and with an eye-hole at the other.

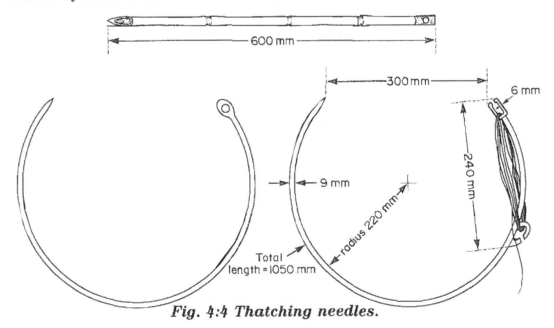

Fig. 4:4 Thatching needles.

If interior access is not simple or if labour is not available the thatcher can use a curved needle. The illustration shows its optimum shape. If wire used for fixing, the needle can be made with extra hooks to carry a long roll of wire — the length needed for each stitch is cut from the needle as required. Figure 4:4 shows various needles.

Screwed Binding This is an adaptation of the stitching method. Instead of looping the stitching wire around the underside of a batten, it is secured to the upper surface of the timber-work by a screw. Screwed bindings are devices made specially for this purpose, comprising a length of stitching wire twisted around a large-head self-tapping wood screw. The screw, 25mm long and preferably a cross-head type, is pushed through the thatch and driven into the underlying timber with a magnetic-tipped telescopic, ratchet-action screwdriver. The two equal lengths of wire are then twisted together over the sway and tightened to provide a secure fixing. Figure 4:5 illustrates the technique.

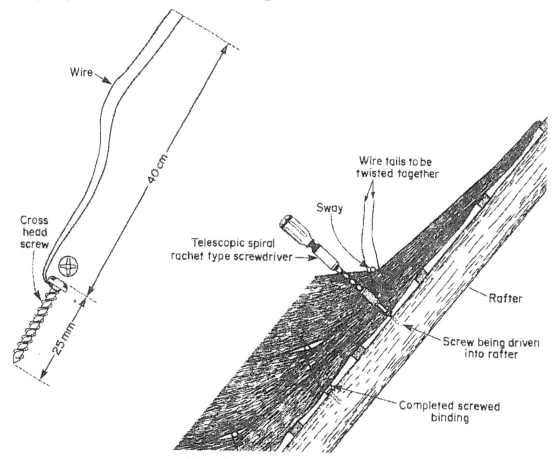

Fig. 4:5 Screwed binding.

This method of fixing the thatch is simple and quick, and has several advantages over the other methods. First, there is no need to have an assistant inside the roof space, as with conventional stitching. Secondly, a ceiling or fire-check lining can be permanently fixed to the underside of rafters and need not be disturbed when the roof is re-thatched. Thirdly, the screwed binding devices are cheaper than steel crooks. Finally, they can also be used for fixing a wrapover type of grass ridge.

Nailing Fixing the thatch by nailing it to the roof structure is only possible if the roof timbers are regularly spaced and can accept the thick thatching nails without splitting. The advantage of nailing is that it is fast, effective and suitable for roofs with constricted interior space. But as several hundred nails are needed for a roof this method may be too expensive, and inappropriate for low-cost building.

Nails, called crooks, should be made to three different lengths: 20cm to suit fixing the eave course of grass, 25cm for all the middle courses, and 30cm for the top course. The number required will depend on the size of the roof. Crooks are made of 8mm diameter steel, sharpened to a point at one end and bent over to form a short hook at the other. The hook secures the sway that is laid horizontally over each course of grass (Fig. 4:6).

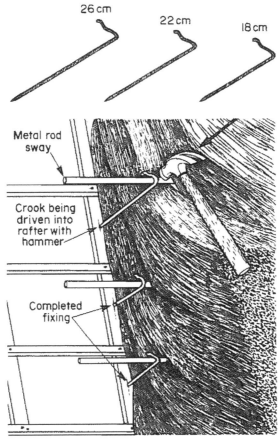

Fig. 4:6 Nailing the thatch.

Sways Sways are lengths of thin wood, or if crooks are used, mild steel 8mm diameter rods. Split bamboo works very well with stitching and screwed bindings. The sway is laid on top of each course and secured to the timber-work, thus holding the thatch in place.

Dressing the Thatch

The leggatt is the distinctive thatcher's tool. Its function is to beat the thatch tightly up into its fixings and to produce a smooth and compact surface. This is called 'dressing the thatch'.

The leggatt is a square, flat block of hardwood, 20cm x 20cm x 3cm thick. To one face is secured a 40cm long handle. The other is cut or drilled to provide a rough surface in which each individual stem of grass is caught whilst the thatcher is beating the roof. The working surface of the leggatt is made either by cutting, with a saw, a series of parallel grooves which are as deep as half the thickness of the leggatt block. Alternatively, and more simply, the surface can be drilled with closely spaced lines of evenly spaced 15mm diameter holes. The drilled leggatt is as effective a tool as the grooved one, though individual thatchers may prefer one to the other (Figure 4:7).

Although only one leggat is essential for good thatching, a professional

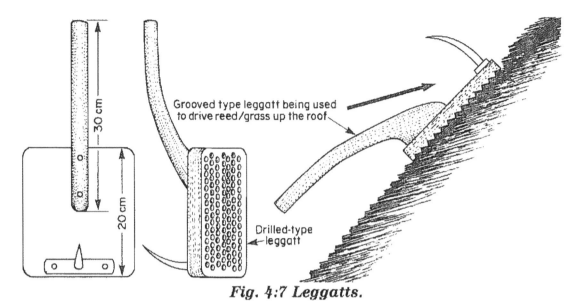

Fig. 4:7 Leggatts.

thatcher may prefer to have several, each slightly different for various parts of the roof and different stages of the work. A heavy leggatt with a long handle is useful for the final levelling and smoothing of the thatch coat. A small leggatt with a curved working suface and a short handle is good for working in the valley junction of two adjoining roof slopes. A short-handle light-weight leggatt might be used for dressing up the work while laying and fixing each armful of grass.

Other Tools and Equipment

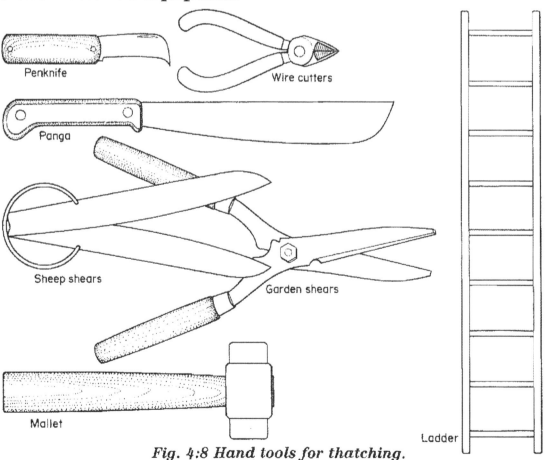

Fig. 4:8 Hand tools for thatching.

A variety of non-specialist hand-tools will also be needed. They are shown in Figure 4:8.

1. A small straight blade knife (a penknife) for cutting the bundle ties and the stitching string/fibres.
2. A mallet for levelling the edge of the eave and gable overhang of the thatch.
3. A large straight-blade knife for trimming the grass at the ridge.
4. A pair of large garden-type shears for trimming the grass beneath the eave to provide a neat and smooth appearance to the roof. Sheep shears/clippers can be used instead.
5. Wire cutters to cut wire if that is to be used for stitching rather than string or fibre.
6. Access equipment is of course obligatory for roof work.

   Ladders or scaffolding are necessary. If scaffolding is erected the working platform should be fixed at 400mm below the top of the wall. The ladder must obviously be long enough to reach to the top of the roof.

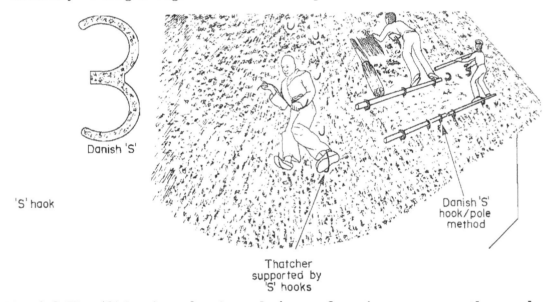

Danish 'S'

'S' hook

Danish 'S' hook/pole method

Thatcher supported by 'S' hooks

Fig. 4:9 The 'S'-hook-and-pole technique of getting access to the work.

Some thatchers prefer to work gradually over the roof in narrow horizontal strips of thatch, completing one strip before moving upwards to do the next section. In this case a simple way of moving sideways along the partially completed roof is appropriate. Two 'S' shaped hooks, one end of which is pushed through the thatch to hook over a batten, support long poles which the thatcher stands on. As one strip of roof is completed the hooks and poles are moved upwards so the thatcher can comfortably reach the next section (Fig. 4:9).

7. The thatcher may need protective clothing, such as gloves, to avoid fine grass splinters, as well as pads to cushion his knees.

Palm Thatching Tools

A palm thatcher will need appropriately shaped knives to split and trim the leaves. He may also need a needle, either curved or straight for stitching the leaves to the roof. And of course he will need ladders or scaffolding to get onto the roof. But there are no tools specifically designed for palm thatching; the durability of a palm-leaf thatch will depend on its thickness and care in laying rather than on the use of specialized tools.

25

5. THATCHING METHODS

Thatching work might best be understood by comparing it with other roof-covering methods. Like tiling, it involves placing the material in ascending and overlapping courses to cover the whole roof surface, so that rainwater flow is directed as rapidly as possible to the ground. In place of a tile the thatcher is working with an armful of grass, or a palm leaf. The thatcher's skill lies in an ability to secure each armful, which may comprise several hundred grass stems, so that each and every grass stem lies parallel to the next one. The join between each armful and each course of grass should be invisible on the surface, and the fixing holding each armful must always be at the same depth within the thatch.

This chapter describes, stage by stage, how this can be achieved. It starts with thatching with rigid stem grasses. Palm leaf and soft-stem grass methods are considered later.

Preparation

Tools, ladders, and materials are assembled on site. The grass must be stacked on the ground conveniently near the work. If necessary, it can be protected from the weather with a tarpaulin. The first task is to ensure that the bundles are the right

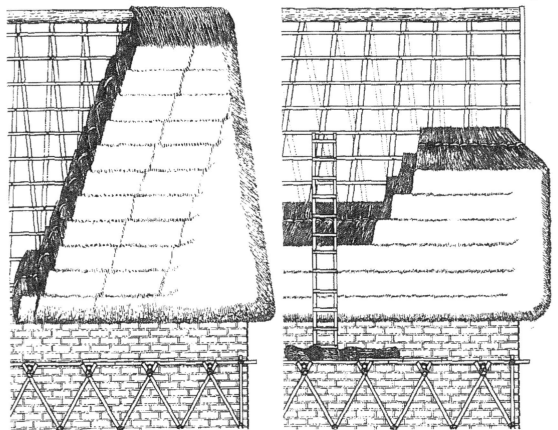

Fig. 5:1 Vertical and horizontal lanes.

size. For most of the roof the standard 550mm circumference bundle will be used, but for the first, eave course of thatch half-size bunches are needed, and these should be prepared in advance. Medium-length 550mm bundles are split into three and each is re-tied with a tight binding located half way up the length of the grass. Approximately eight of these bunches will be needed for each metre length of eave course.

The rest of the bundles are then graded and stacked separately in three categories: (a) long, (b) short, and (c) coarse (bent and twisted stems and those without much taper). Medium-length and long grass is used for work on the level sections of roof. Short is best kept for the topmost layer and for working at an angle up the gable. Coarse grass can be used for backfilling and for making ridge rolls.

Before starting, the thatcher should decide whether to work in vertical or horizontal sections over the roof (**Fig. 5:1**). This decision will be partly personal preference, but it also depends on the roof shape and ease of access. Working vertically, from a ladder, the thatcher will fix the grass in ascending courses each about one metre wide. He will complete an entire section, or lane, up to the ridge before starting an adjacent lane. A tall thatcher may comfortably work a slightly wider lane.

For horizontal thatching the roof would be divided into strips of three or four courses of thatch. The first strip would be thatched from a ladder or scaffolding, the platform set 50cm below the eave level. Thereafter the thatcher gains a foothold on the roof by standing on poles which are supported at each end on 'S'-shaped brackets hooked around the roof battens (Fig.4:9). As each horizontal section is completed further brackets are inserted through the uppermost layer of thatch to carry the poles which the thatcher will stand on to do the next section. The poles and brackets are removed when the roof is completed.

Thatching — Step by Step

Assume here that the roof is a gable type, with a linear ridge and that work is being done in vertical lanes from a ladder which may be standing on the ground or on the scaffold paltform. Assume also that the thatch is being stitched to the battens, although the same technique applies for the other fixing methods.

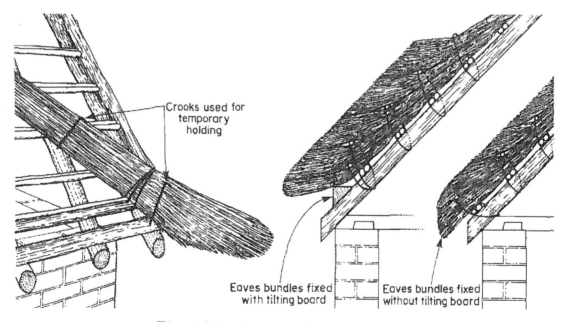

Fig. 5:2 Fixing the first eaves bunch.

28

1. Thatching begins with the eave course at the bottom right-hand corner of the roof. (A left-handed thatcher would start from the other end). This eave course must be very securely fixed as the grass overhangs the wall and is thus vulnerable to wind damage. It is also the foundation course and will provide the support for subsequent courses of thatch.

The first eave bunch is laid at 45° on the angle between the eave and the gable, overhanging the wall by some 300mm. It is tied tightly to the first batten and should be visibly compressed and bowed by the eaves tilting board (Fig.5:2). Subsequent eaves bunches are turned away from the angle and fixed down, until, within 1.5m, they lie parallel to the rafters (Fig. 5:3).

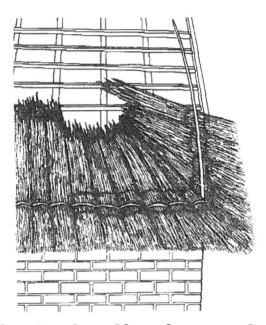

Fig. 5:3 Forming the gable and eave overhang.

2. Eaves size bundles (see section 5:1 above) are then laid up the length of the gable at the same 45° angle and stitched to the gable rafter. The eaves bunches are then wedged tightly into their fixing by tapping the butt ends of the grass with the leggatt, thus forming the precise line and shape of the eave and gable overhang.

3. Having completed the gable foundation course, and the first metre width of the eave, a thin layer of coarse grade grass is spread over the structure of the roof that will be covered by the first lane of thatch. This backfilling grass prevents the ears of each course being driven and buckled beneath the battens.

4. The first course of the surface coat thatching is begun by laying an armful of grass over the eaves course, making sure that the butts are level, and building up to the full 300mm thatch thickness. The grass is dressed to match the roof pitch by tapping the butts upwards with a leggatt. This first armful is then secured to the structure with a horizontal sway laid about 600mm from the butt end of the grass and held with stitch to the first roof batten. Second and subsequent bundles are laid to the width of the lane, making sure that each merges tightly with the edge of the previous bundle and that the grass stems lie parallel to those of the eaves bunches. The grass is then all dressed into place and held by a stitch, every 300mm, which passes through the grass around the batten and over the sway.

Finally the grass of this first course is driven tightly into the fixing with the leggatt so that the butt ends form a pitch identical to that of the rafters. A slight lip is left at the top of this, and each subsequent course; this lip will be driven up with the grass of the next course thus forming a neat and invisible junction (Fig. 5:4).

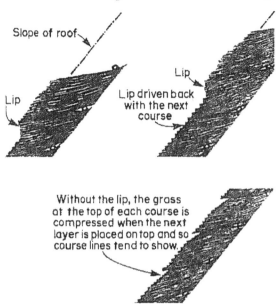

Fig. 5:4 Leave a small lip at the top of the course.

5. Work proceeds course by course to the roof apex, particular care being taken to ensure that each sway is at the same depth within the thickness of the thatch, and that the total thickness of thatch is maintained at 300mm throughout. The edge of

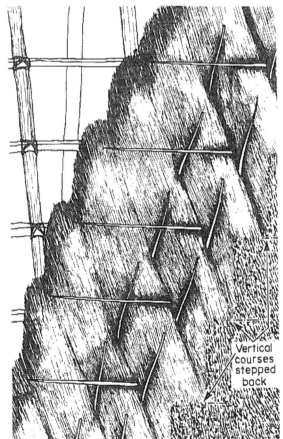

Fig. 5:5 Avoiding vertical joins between lanes.

each course is supported temporarily with two pointed pegs driven into the previous course. These are moved sideways as the work of completing each course progresses, lane by lane, until the whole roof is covered.

Within each lane, every course ends slightly back from the edge of the previous, lower course rather than directly in line with it. Thus the joins between each lane are stepped (Fig. 5:5). This prevents the formation of vertical surface lines and avoids the possibility of guttering as the thatch decays.

6. Lane by lane the roof is thatched until the entire surface is covered. Valleys, hips and other features are worked by building up with extra grass and beating the surface to form smooth sweeping curves rather than sharp angles, so that rainwater flow is evenly dispersed.

7. Towards the top, shorter bundles of grass are used. A ridge-roll of 150mm diameter and as long as the length of the ridge, made up of coarse grass lashed every 150mm. is then secured to the ridge board (Fig. 5:6). This roll is needed to maintain the steep pitch of the top two courses of grass. The uppermost course is fixed to the top batten, and the grass that oversails the apex is cut off level with the ridge roll.

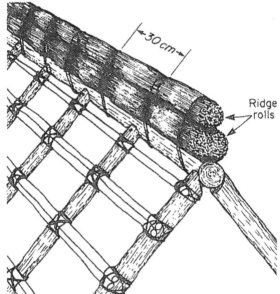

Fig. 5:6 The ridge roll.

8. After both sides of the roof are covered, the ridge must be sealed. The purpose of the ridge is to cover and protect the exposed grass and fixings of the top course of thatch. Various materials can be used. Figure 5:7 shows several different ridging methods. If grass is used it must be flexible enough to be bent over the roof apex without breaking. Soaking it may make it flexible. A grass ridge is secured by stitching, but it is advisable to use non-corrosive metal wire for maximum durability. Alternatively, the ridge may be made of metal sheet, held by stitching to the upper batten through holes punched in the lower edge. A cement mortar capping, reinforced with wire mesh (ferrocement) may also serve as the ridge.

9. Finally, to eradicate marks left in the surface by the ladder, and to produce a smooth and compact thickness of thatch the roof should be beaten hard with the leggatt. The stiffness and taper of the grass means that each stem and each course will be wedged tightly into the fixing with this final beating.

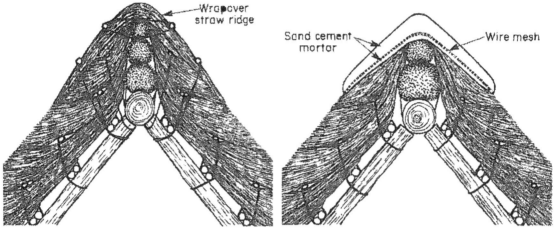

Fig. 5:7 Ridging.

Soft-Stem Grass Thatching

Thatching with flexible soft-stem grass is similar to rigid-stem thatching, although it is not so durable. The technique differs because it is not possible to use a leggatt to wedge the grass tightly into its fixings. The shape of the roof and the compactness of the thatch depends on the thatcher's experience of laying a consistent coat with unsympathetic material.

The neat appearance of a newly thatched soft-stem roof is created by carefully trimming the overhanging eaves and gables, and by combing the surface with a small hand-held rake. This will remove loose grass and, in aligning the stems parallel to each other, create a more compact coat of thatch.

Palm Leaf Thatching

Palm leaves are naturally occurring pre-formed panels of thatch. They are fixed in quite a different way to grasses and, depending on the species, will need careful preparation before being fixed to the roof. The techniques are described in the following two sections.

Feather-type leaves

Palms, such as coconut palms, which are composed of leaflets attached to both sides of a linear stem are called feather-type leaves. The leaves are generally cut from the trees when mature (or when needed) and are left to dry in the sun. Then the central rib is split so the leaflets hang from one side, thus making two shingles from each leaf. Alternatively the leaflets may be bent all to one side of the rib and perhaps plaited together to strengthen the thatch panel (Fig. 5:8). If the leaflets are attached at a sharp angle to the central rib it may be necessary to cut them off and make them into shingles. Those of the *nipa* palm are prepared by folding the detached segments down one-quarter their length over a firm support such as a strip of bamboo so they overlap, and sewing them down firmly. These are called 'ataps' and are usually about 750mm long. 'Ataps' are fixed to a roof in the same way as feather-type leaves.

Work on the roof proceeds much like shingling, starting at the bottom by tying the rib to the roof rafters. Battens are not needed as the rib of each leaf serves this purpose. The thickness of the thatch will depend on the gap between each rib and the one above it, but the thicker the thatch the more durable it is likely to be. However, care has to be taken to ensure that the leaflets are sloping steeply downwards — if the thatch is made too thick it is possible that they will lie almost

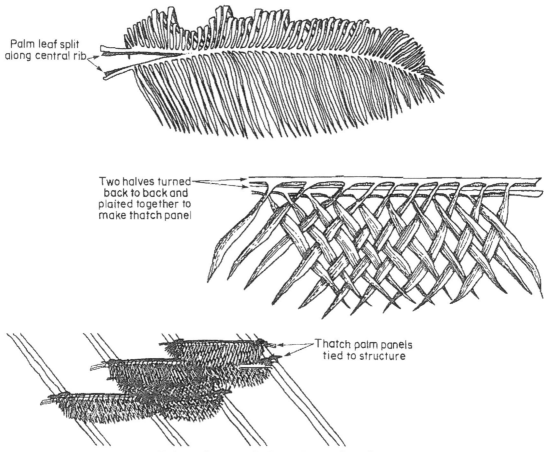

Palm leaf split along central rib

Two halves turned back to back and plaited together to make thatch panel

Thatch palm panels tied to structure

Fig. 5:8 Palm leaf thatch — feather-type.

flat although the roof profile shape is sloping — and water may then be carried deep into the thatch along each leaf blade.

The major drawback to palm thatch on new gable-walled buildings is the difficulty of making a strong and watertight gable. With grass, each layer is turned through 45° to make an overhang, and the fixings of each course are covered and protected; but this is not possible with palm leaves. Thus at each end of the roof the tip of each rib will be exposed and liable to direct water into the thatch coat.

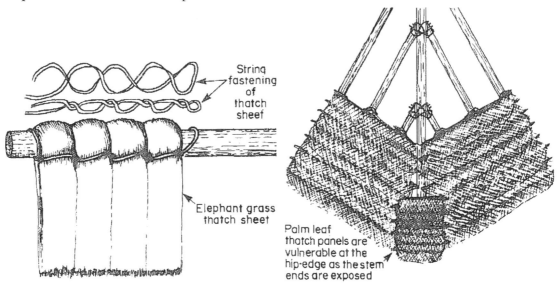

String fastening of thatch sheet

Elephant grass thatch sheet

Palm leaf thatch panels are vulnerable at the hip-edge as the stem ends are exposed

Fig. 5:9 Thatching a hip-edge with palm leaves.

For this reason, palm leaves are traditionally only used on full hipped roofs — where the eave line is continuous and unbroken around all four sides of the building. But this presents the problem of making a secure and weather-tight thatch up the length of each of the four hip rafters. If the thatch ribs are flexible they can be bent around the hip, but because the leaflets will naturally hang downwards the vulerable hip edge will only be thinly covered and prone to leakage. Alternatively, shortened lengths of leaf panel (about 300mm long) may be fixed up the line of the hip after adjacent slopes of the roof have been covered. However, the ends of each hip panel rib will be exposed, as they are on the edge of gable-type roofs (Fig. 5:9).

Fan-shaped Leaves
This type of leaf needs no preparation other than drying and flattening. They are then tied or otherwise fastened (Fig. 5.10) to the roof framework like shingles with the leaf blades pointing downwards.

A roof thatched with these leaves has the same weakness at hips and gable verges as a feather-type leaf roof.

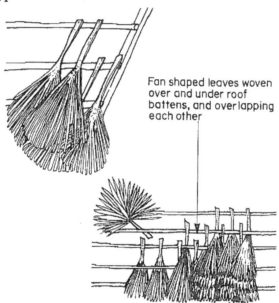

Fan shaped leaves woven over and under roof battens, and overlapping each other

Fig. 5:10 Fan-shaped palm leaves.

The ridging of palm-thatched roofs is similar to grass thatch ridging. Use either flexible leaves, bent over the apex and secured by stitching, or metal sheet or ferrocement. The purpose is to direct rainwater down the outside surface of the thatch and to prevent it getting into the body of the thatch.

Conclusion

Rigid-stem thatching is the most durable type, particularly if the grass meets all the specifications set out in Chapter 3. Durability in temperate regions may exceed 60 years providing that the ridge remains watertight or is regularly repaired. In the humid tropics where conditions favour fungal decay, life expectancy may be much reduced but should nonetheless exceed 30 years.

Palm leaf and soft stem grass thatching will not last more than 15 years and probably significantly less than 10 years.

6. MAINTENANCE AND REPAIR

All buildings need regular maintenance. If they are skilfully built they should need less attention and this is true for thatch as it is for all other work. Thatch does however decay and will eventually reach a point at which complete renewal is unavoidable. Maintenance cannot prevent this, but some jobs do need to be done regularly. This chapter lists these and offers a typical maintenance plan.

Allowance must also be made for unexpected damage. This chapter considers which problems are most likely and how to repair the roof if they do occur. Finally it explains how to identify a worn-out thatch before it leaks.

Maintenance

The ridge should be the only part of the roof that needs regular attention and it is important that it is maintained. If water penetrates through the ridge it will certainly cause decay of the top course and may eventually soak through the rest of the thatch. Once this happens, the fixings that hold each course can be corroded and weakened.

A straw or grass ridge that is made by the wrapover method and stitched or pegged to the roof should be renewed every eight or ten years, and perhaps more frequently. Other ridging types, such as ferrocement or sheet metal may last longer, and perhaps even as long as the thatch coat. Nonetheless, all ridges should be inspected regularly and this applies equally to a roof coated with palm leaf thatch.

Re-ridging is simply a matter of stripping off the remains of the old ridge and replacing it with new and tightly secured material. The thatcher doing the work will have to take particular care to avoid damaging the surface of the thatch, which after ten or so years' exposure will have become brittle. In particular the ladder must be placed so that it lies flat on the roof; do not let it dig into the eave overhang.

Regular Inspection

(a) It is possible that the thatch may have slipped slightly as the material has settled. This should be clearly visible within a year of thatching as faint horizontal lines will show on the roof where the courses have separated. The remedy is to go over the roof surface tapping up all the grass with a leggatt, thus wedging it tightly back into its fixings. However, this should only be done if it is absolutely necessary, as even a light dressing will damage the tips of the grass stems and encourage more rapid decay.

(b) The thatch surface must be kept clear of vegetation which can impede rapid rainwater drainage. In some parts of the world, people train gourd-bearing creepers over the roof; this should not be encouraged. Similarly, water which drips from overhanging trees will hasten decay, so the offending branches should be pruned. However, remember that thatch which is shaded tends to last better because sunlight is one of the causes of decay, so only the directly overhanging branches should be cut back.

(c) Remove moss and other lichen if it grows on the roof. To avoid disturbing the thatch unduly, the best treatment is to spray the surface with a suitable herbicide. The vegetation will then wither and eventually disappear with natural weathering. Copper sulphate solution is generally effective, though spraying may have to be repeated every few years where perpetually damp conditions favour this sort of growth. Alternatively, a length of thick copper wire fixed along the lower edge of the ridge effectively discourages fungal growth. Whenever it rains, diluted carbonates and hydroxides of copper soak the roof and kill off vegetation.

Repairs

Storm damage and bird damage are the two problems most likely to need attention.

Birds, particularly during the nesting season, may either pull out thatch or actually build their nests under the eaves. Nests should be removed, though preferably not until fledglings have flown, and holes filled. This is simply done, provided that the hole is easily reached. Make a bundle of thatching grass that will fit tightly into the hole. Tie it with two bindings close to the ears of the grass and insert a short (40 cm) sharpened stick beneath the ties (Fig.6:1). This bundle can be pushed up into the hole and the loose ends trimmed off with clippers to produce an almost invisible and permanent repair.

Fig. 6:1 Repairing bird damage.

If the damage is particularly severe the roof can be covered with wire mesh, but this is expensive and could itself cause more damage than birds. The mesh catches leaves and debris and impedes water flow. Use either 20 or 22 gauge 19mm wire mesh in metre-wide rolls. The best method is to roll the mesh down each side of the roof and to join it along the ridge. Each drop is then joined to the next by twisting the wire together using a metal hook (Fig.6:2). Finally, nail it to the top of the wall or tuck it under the eaves.

Storm damage, usually the effect of strong wind, is most likely either at the ridge or

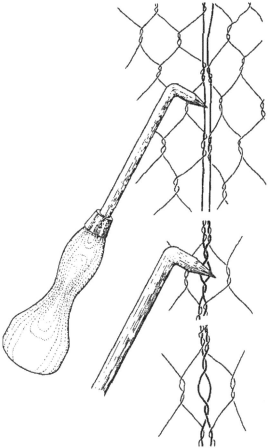

Fig. 6:2 Fixing wire mesh to a roof.

the eave. Ridge repairs are straightforward and are simply a case of renewing the damaged section. If, however, thatch has been stripped from the eave, repair work may affect the thatch above and the thatcher may then have to replace a lane of roof right up to and including the ridge.

Always start a repair at the lowest point of damage. If only a handful or two of grass is needed. This may be replaced in the same way as described above for bird damage, but a larger hole may involve rethatching a section of the roof. It is important that the join between the original thatch and the repair is tight and flush so that the natural erosion of the roof surface will not be accelerated by repair work. See Figure 5:7. To match the partly worn condition of the old roof it may be necessary to select very short grass for the repair.

When to Rethatch

The lifespan of thatch depends on the quality of the material and the skill of the thatcher, as well as climate and other local conditions. It is more difficult to predict exactly when it will need replacing. Palm-leaf thatch will rarely last more than five years, but skilfully-laid, rigid-stem grass thatch should remain weather-tight for at least 20 years and perhaps more than twice that long. Whatever the anticipated durability, every roof should be inspected so that re-thatching can be organized before it leaks.

Thatch is worn out when it has decayed to the point when the fixings of each course become exposed on the surface. When this happens, water can be channelled through the thatch and into the building by running down the stitching ties or thatching nails. The roof, or the worn out section of it, should be rethatched *before* this happens.

The sure sign that thatch needs replacing is the appearance of horizontal lines showing on the surface of the roof, often accompanied by slippage of one or two areas of thatch. Within a few years, the course fixings will be exposed and leaks will occur.

Ideally, the entire thatch coat will gradually decay at the same rate, but this is rarely the case. Usually, one part of the roof will be worn out before the rest. In the northern hemisphere, the south-facing side has a shorter life than the shaded north-facing side, so one side will often need re-thatching first. The reverse may apply south of the equator, but in the tropics differential decay rates may be negligible and both sides of a roof will usually need re-coating at the same time. Additionally, certain parts of a roof may wear out faster than the rest, particularly in valleys, beneath chimneys or where the roof pitch is shallow over windows. Small areas such as this can be repaired, as storm damage should be, but more extensive decay warrants complete renewal of at least one side of the roof.

Although it is uneconomic to replace thatch which is not worn out, it may be worth sacrificing a section of good roof so as to avoid the inconvenience of having to employ a thatcher again within a year or so.

Re-thatching

Re-thatching means fixing a new, 300mm-thick coat of thatch. Like any major building work, this will be disruptive, particularly so in the case of thatch, as a large quantity of semi-rotten grass has to be removed.

Usually, the job involves stripping and disposing of the old thatch, repairing the roof structure if necessary and then attaching the new thatch in the same way as the original covering.

Alternatively, if the original thatch is still securely fixed, it may be possible to lay this new thatch on top of the remains of the old coat. The advantage of this second approach is that the insulation value provided by the remaining old thatch is retained, and there is a possible financial saving through not having to pay the thatcher to strip and dispose of the old thatch. But this recoating can only be done with the softer and more pliable types of rigid stem grasses which can accommodate the softer underlay of old grass. Water reed, for example, is too stiff and must be laid on a firm base of rigid timber-work.

7. PROBLEMS WITH THATCH

Thatched roofs have two serious physical disadvantages — fire risk and a tendency to harbour insects and other pests. How can these problems be minimized?

Fire

Grasses and palm leaves are combustible, particularly when they are dry. Apart from sandwiching the thatch between two incombustible layers there is no way of making them completely fire proof; but various techniques may be used to guard against the most likely sources of fire.

Building Design

Common sense is the most valuable precaution against fire, and this applies both to the design of the building and to its subsequent occupation. The design must exclude unnecessary fire hazards. Electrical wiring should always be carried out by professional electricians, particular care being given to any installation in the roof space. Chimney design and construction must isolate hot flue discharges and must

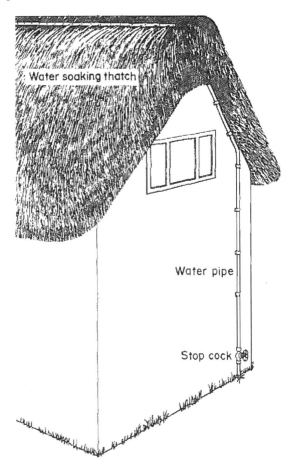

Fig. 7:1 Sparge pipe.

release them at least a metre above the top of the thatch. Cooking arrangements must minimize the chance of accidental fire. Buildings should be suitably protected with lightning conductors, and should preferably also have an outside tap permanently fitted with a long hose pipe. These are the obvious precautions, and they must be complemented by fire-conscious household activity. For example, fires should never be lit close to the building, and always downwind of it.

Surface Coating
Covering the thatch with a coating of non-combustible material protects against external fire risks. Indian research proved this using either sand/cement/lime slurry or bitumen-stabilized earth plaster. But these, and other possible coverings such as fibre glass matting will add to the expense and weight of the thatch, and of course will completely obscure the aesthetic appeal of a well-thatched roof. However their major drawback is that the coating will not have the same expansion-contraction character of the thatch beneath, so cracking will inevitably occur. This will let in water which will cause rapid decay of the unventilated thatch.

Chemical Treatment
A great deal of research and experimentation has gone into the search for a chemical treatment to make thatch fireproof. So far none is successful except in the short term as all the chemical fire retardants are water soluble. Whether applied by dipping the material before it is placed on the roof, or by spraying the thatch, it has been shown that the chemicals on the surface are washed off within a year, and within the thatch coat humidity changes eventually break the bond between the chemical and the thatch, rendering it useless. Treatment based on commonly available chemicals have the added and serious disadvantage of encouraging the growth of mould and surface algae. Tests in Britain found that a year-old treated roof showed as much decay as an untreated ten-year-old roofs. Despite, and perhaps because of these drawbacks, further research is being carried out, but so far there are no convincing results.

However, a thin surface coating using a waterproofing agent may be useful as a decay retardent. In India a cheaply available by-product of the cashew nut processing industry was found to increase the life of palm thatch which has been painted with the liquid. Wherever a non-water-soluble liquid is cheaply available in sufficient quantity it may be used to protect thatch against decay.

Incombustible Underlay
A fire needs a continuing oxygen supply. For this reason a partial reduction in combustibility can be achieved by fixing a fire-resistant lining underneath the thatch. Additionally, and perhaps of more importance, is the fact that many house fires start inside the building — so lining the thatch will protect it from internal fire sources. In Europe this has been recognized by risk assessors who advise insurance companies, and thatched buildings which have been lined now attract lower premium payments. The lining does not prevent thatch being ignited from an external source, but it does reduce the speed the fire spreads so giving extra time to extinguish the fire.

A fire check lining is usually attached to the rafters after thatching work is complete. It may comprise sheets of fire-resistant plasterboard tacked to the timbers. If this is not available, it is possible to produce a similar degree of protection by fixing rough cloth — such as hessian — which can then be coated with a sand/cement/lime slurry. The main disadvantage with an underlay is that it may have to be removed when re-thatching becomes necessary, particularly if the thatch is fixed by stitching to the rafters.

Sparge Pipes
Where a sufficiently high-pressure water supply is available, perforated metal pipes may be fitted to the top of each side of the roof and controlled by stop taps at ground level. If there is a risk of fire the water can be turned on to soak the roof. The pipes and supply should be capable of providing 2.25 litres per metre^2 of roof per minute. This exceeds the capacity of a normal domestic water supply, so special arrangements will have to be made.

Thatch and Fire Risk — Conclusion
Perhaps the most effective protection against fire is high quality thatching. Whereas a loosely laid thin coat of rough thatch may be compared to a sheet of paper, a tightly thatched, thick and compact thatch is more like a book. Paper burns readily but books are hard to ignite. The design guidelines offered in Chapter 3 should also be followed, and regular maintenance will keep the roof in weather tight and less fire-prone condition.

Should fire break out-two sensible precautions may minimize the damage. First, a plentiful supply of water should always be available. Ideally, there should be an outside tap with a long hosepipe permanently attached to it. Secondly, a long-handled metal rake should be available for pulling down smouldering thatch and creating a fire-break.

Insects and Other Pests
Thatch does often provide a home to insects, and occasionally a source of food for rodents or nest material for birds.

To protect the occupants from the nuisance of insects the best protection is a ceiling. This may also double as the fire check lining described above or may be the more usual horizontal ceiling. In the latter case particular care should be taken to ensure that the junction between wall and ceiling is well sealed. This will also prevent dust from the roof entering the rooms below.

Bird damage has been dealt with in Chapter 6. A further precaution which sometimes works is to fix an artificial but life-like bird-of-prey in a prominent position on the roof. Rodents and occasionally animals can also damage thatch.

There is no universal answer to any of these problems other than making sure that there is no nutrient value, particularly grain seed in the case of cereal straw, in the thatch.

8. SPECIFICATION FOR THATCH

There are simple methods of measuring a roof so as to calculate how much thatch is needed and how long it might take to thatch a roof.

Measurement

The roof is measured to give the area of timber-work between the eaves and ridge and from one side of the roof to the other. This is called the 'superficial roof area' and is shown in square metres. In addition, the length of thatch overhanging the eaves and gables and the thicknesss of the thatch must also be included. Experience has shown that adding 10 per cent to the superficial roof area produces a sufficiently accurate estimate of thatch area.

Thus a simple, double-pitched roof with an eave length of 5m and rafter length of 4m will have a superficial area of $20 \times 2 = 40m^2$ and therefore a total thatch area of $44m^2$.

If the roof plan is complicated, for example by chimneys or roof windows, the thatch area is adjusted as necessary. However, although less thatching material may be needed because of these features, they are difficult and time consuming to thatch. A cost estimate must therefore include an additional charge, based on the local labour rate, for each roof window, valley, hip and chimney.

The other information needed for a cost and materials estimate is the ridge length, measured in metres.

A sufficiently accurate figure for the area of a circular roof is found by multiplying the distance from the apex down over the eaves, inwards to the wall, by half the circumference.

Materials Estimating

Grass Thatch Ten bundles of grass are needed for each square metre of roof. This applies to rigid-stem grass with an average length of 1.5m, laid 300mm thick on a roof pitch of 45°. Bundles are assumed to be 55cm in circumference around a binding which is 30cm from the cut end, as this is the size that is most easily handled by the thatcher on a roof.

The figure will of course vary according to the type and length of the grass, the thickness of the new thatch coat and the pitch of the roof. Less grass is used if the pitch is steeper, or if it laid thinly, or if the grass is particularly long, for example, greater than 2.2m.

In addition to grass, fixing materials are also needed. On average, each square metre of thatch will be secured at ten separate points, requiring approximately 6m of stitching wire or rope, or ten thatching nails, as well as 4m of sway.

Thatching with soft-stem grass consumes a similar volume of material.

The ridging estimate depends on the type of covering selected. A linear metre (covering both sides of the roof) of wrapover ridge made of soft-stem grass uses about 5 bundles which are held at 20 fixing points with two, metre-long sways on each side of the roof.

Ferrocement ridging, extending 600mm from the apex on each side and 75mm thick uses 0.09m³ of 1:4 cement/sand mortar and 1.2m² of galvanized wire mesh per linear metre.

Weight of Thatch In order to calculate the load that the roof timberwork will support, the weight of the thatch must be known. The heaviest material (water reed) laid to 300mm thickness does not exceed 45kg/m². Wheat straw is lighter, averaging 25kg/m². Roof timbering must of course be capable of supporting the extra weight when thatch is wet — an additional load of 10 kg/m².

Palm Thatch A general specification for palm thatch cannot be provided, as each type of palm leaf and the way it is used is so different from the others. However, the guidelines of roof pitch and thatch thickness that apply to grass thatch are also true for palm thatch. Thus palm leaf thatch will last best if it is laid thickly on a steep roof pitch (exceeding 45°).

Work Rate

An experienced thatcher, with an assistant, should be able to complete between 8 and 10 square metres of plain grass thatching during an average eight hour working day. Working around chimneys, roof windows, valleys or hips is more time consuming.

Ridging rate will depend on the covering used. Five metres of the wrapover grass type should take about a day for a two-man team; other types should be completed more quickly.

Finally, the work rate will also depend on the height and the ease of access to all parts of the roof. Scaffolding, though more expensive than a ladder, is easier and safer to work from.

44

GLOSSARY

Backfill — A thin layer of grass laid on top of battens to allow courses of thatch to slide up the roof frame without catching on the timberwork.

Battens — Thin strips of sawn softwood (25 x 50mm) fixed horizontally to the *rafters*. Split bamboo may be used if the thatch is stitched to the structure.

Bunch — Small bundles of tightly tied grass used for the *eaves* course of thatch. Three bunches may be made from one *bundle*.

Bundle — A specific quantity of grass. A bundle is 55cm in circumference when measured around the binding, which is 30cm from the *butt end*.

Butt end — The lower, cut end of a grass stem.

Course — Strips of thatch, usually about 150mm deep, laid horizontally along the roof.

Crook — Steel nails, used in conjunction with *sways* to secure *courses* of thatch.

Culm — The clean stem of grass from which soft leafy material has been removed by combing.

Depth gauge — A thin stick marked off in centimeters and millimeters used to check that the thatch is being laid to the desired thickness.

Dressing — Tapping the *butt ends* of grass upwards with a *leggatt* to produce the correct surface slope.

Ear — The seed-head end, or tip of a grass stem.

Eave — The bottom edge of the roof which overhangs and protects the top of the walls.

Flashing — Covering strip used to seal the join between thatch and an adjoining surface, such as the side of a chimney. Preferably made of flexible metal, such as copper or lead.

Gable — The finished edge of thatch overhanging the side walls of a building.

Hip — The outstanding edge formed by the meeting of two roof surfaces which do not finish with a *gable*.

Lane — A section of thatch from the eave to the ridge, about one metre wide.

Leggatt — A wooden tool, shaped like a bat with a grooved or drilled out surface used to dress the grass stems into place.

Peg — 30cm-long thin pointed sticks used to temporarily hold the edge of the *course* whilst laying the thatch. Two are used for each course. They are discarded as each course is completed.

Rafter — Sloping timber extending from the eave to the ridge: the primary support of a narrow span roof structure.

Ridge — The apex of a double pitched roof.

Ridge roll — Bundle of tightly tied grass, 150 mm in diameter and as long as the ridge. Needed to maintain the steep slope of the upper two courses of thatch.

'S' Bracket — Curved metal 's'-shaped hooks used to support each end of long poles which thatchers stand on to get access to the work. (Used only if the work is being done in horizontal rather than vertical *lanes*).

Screwed Binding A fixing device comprising a length of wire twisted around a large-head self-tapping wood screw. The screw is preferably a cross-head type, to be driven into the batten with a magnetic tip pump-action screwdriver.

Side rake — A hand-held tool, like a comb, used to clean grass bundles and to smooth the surface of soft-stem thatch.

Sparge pipe — Perforated metal tubing fixed horizontally along the *ridge*, connected to a high-pressure water supply and controlled by a stop tap.

Sway — Thin lengths of wood — often split bamboo — laid on top of each *course* and secured by a stitch, *crook* or *screwed binding*.

Tilting board — A length of timber fixed along the bottom of the roof structure which is needed to create a firm base for the eave course of thatch. It must be some 50mm thicker than the subsequent roof *battens*.

Valley — An intersection of two sloping surfaces of roof — the opposite of a *hip*.

www.ingramcontent.com/pod-product-compliance
Lightning Source LLC
Jackson TN
JSHW040739140125
77033JS00047B/1011